EXPLORE OUR NATIONAL PARKS

MR. RINGTAIL'S
ZION ADVENTURE

BY DR. MIKE KOZUCH

my ScienceBlast for Kids

Title: Mr. Ringtail's Zion Adventure
Author and Photographer: Michael Kozuch
Publisher: myScienceBlast
Imprint: myScienceBlast for Kids
Website: myscienceblast.com

ISBN: 979-8-9936955-1-8
Printed in the United States of America

Photo Credits:
Photos © Michael Kozuch except where noted.
Ringtail image courtesy of the National Park Service/NP Gallery (public domain).

First Edition published 2026
Revised 2026

Dedicated to all the young explorers who
keep asking "why?"

SHALL WE BEGIN ?

Hi there, explorers! My name is **Mr. Ringtail**, and I'll be your canyon guide. Zion is full of cliffs with rocks that formed when the dinosaurs walked the earth!

You might be wondering what a **ringtail** is. I am a raccoon-like animal that mainly comes out at night, so you probably won't be able to see me unless you are very lucky. I am also pretty good at climbing around the rocks and cliff ledges.

Are you ready to start on our geological adventure?

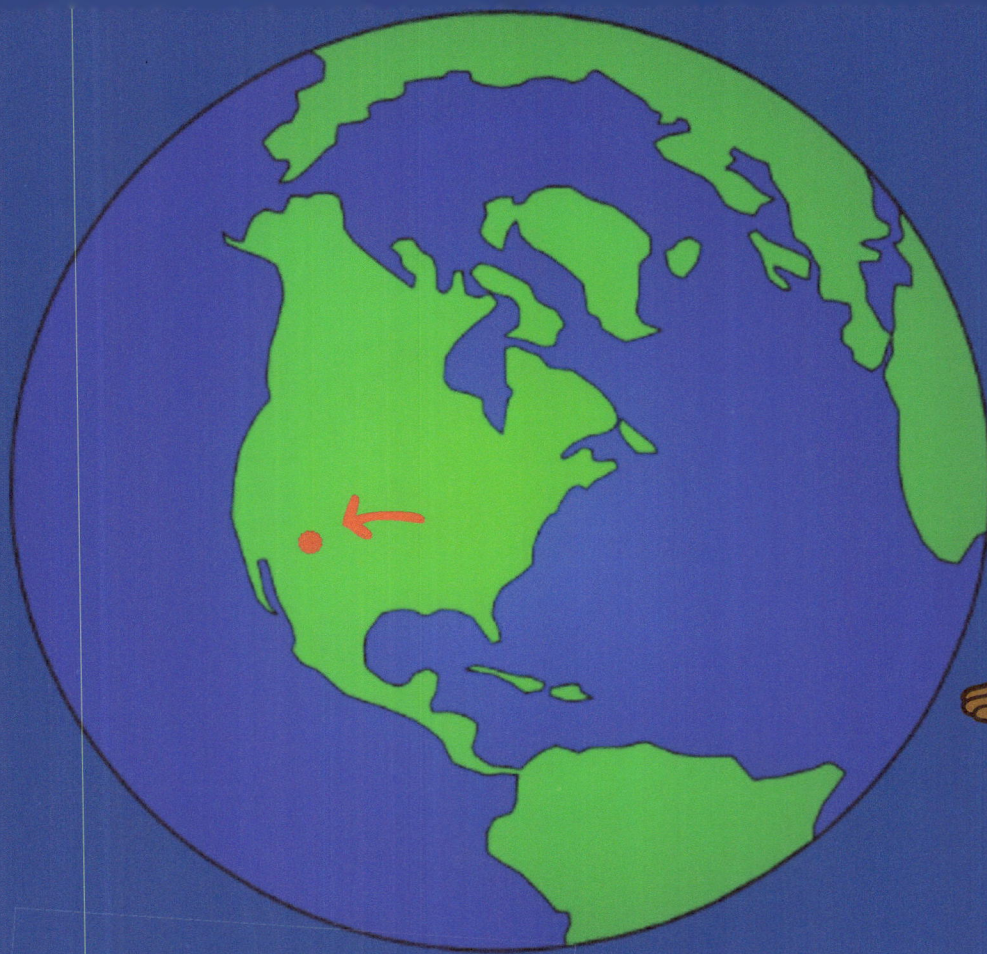

Some of you may be asking, "Where is Zion?"

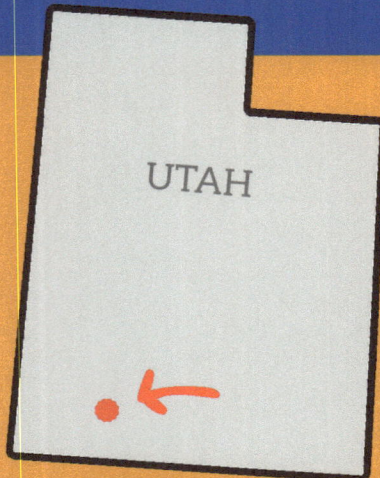

UTAH

Zion National Park
lies in southwestern Utah. It was the first national park in Utah and was officially named a national park in 1890.

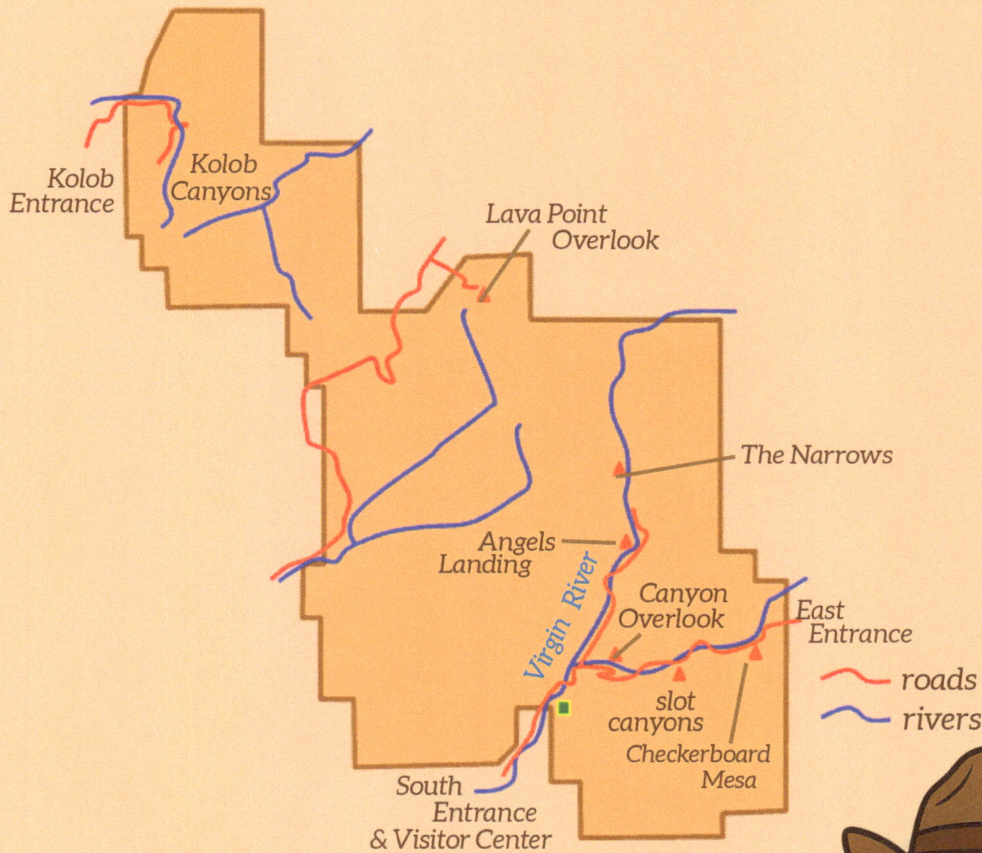

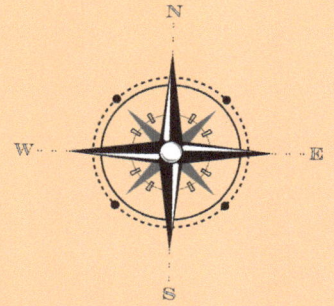

Zion National Park

Kolob Entrance

Kolob Canyons

Lava Point Overlook

The Narrows

Angels Landing

Virgin River

Canyon Overlook

East Entrance

slot canyons

Checkerboard Mesa

South Entrance & Visitor Center

N
W · · · E
S

~ roads
~ rivers

It's always a good idea for young explorers to know where they are going. You can use this map to keep track of all the places we'll talk about while visiting the park.

About 200 million years ago this area was a giant **desert** full of enormous sand dunes. Eventually, this desert got buried by other rocks and sediments. The weight of all that material turned this sandy desert into a rock called **sandstone**.

Here is a closeup of those sand grains made of a mineral called **quartz**. Notice how round they are. That is because they rolled around with the wind in ancient deserts. **Quartz** is usually a grayish color but here you can see the grains are covered in a rusty iron mineral called **hematite**. This is what gives many of the cliffs a reddish color.

See how the bottom of the cliffs are red but the tops are white? At one time all the sand grains were red because of that iron coating. Over time, water washed away some of the iron, leaving the top part white. Same **sandstone** rock, but different colors!

Geologists call this group of rocks the **Navajo Formation.**

Scan here to see more about the Navajo Formation!

See the dark streaks and stains on the walls? That's called **desert varnish**! It's a thin coating of minerals that have **iron and manganese.** Tiny **bacteria,** microscopic organisms, take these minerals from dust and cement them to the rock. It can take thousands of years to make those streaks!

Many of the cliffs in Zion have crisscross wave-like patterns. That's because these are the surfaces of those ancient sand dunes. These patterns tell geologists which direction the the wind was blowing and how it moved the dunes. They call this **crossbedding.**

Scan here to see more about ancient sand dunes!

This is Checkerboard Mesa

This curious hill has vertical and horizontal cracks. The **horizontal** ones are made by the crossbeds, but the **vertical cracks** are made from the changes in temperature. Hot days and cold nights make the rocks stretch and then shrink, just enough to crack.

The view from Lookout Point.

This amazing view reminds us about the power of rivers. If you give them enough time, they can create deep valleys. Rivers in Zion began changing the land starting 18 million years ago.

The **Virgin River** is the main river that cuts through Zion and this canyon is known as the **Narrows.**

Scan here to see how rivers formed these canyons!

We've covered a lot! Let's see how much you can remember.

1. What do we call the rock made of sand grains from a desert?
2. What is the name of the formation of the red and white rocks?
3. What are the criss-cross patterns in the rocks called?
4. What are the dark stains on the cliffs called?
5. What do you think Zion will look like in a million years?

Zion Geology Word Search

Here is your chance to see if you can spot all the new words you learned in a Word Search. Circle the groups of letters that spell the words you find below the table. The words can go across, down, or even up (a bit trickier).

Photocopy this page and give it a go!

D	Q	L	A	C	I	V	A	R	N	I	S	H	D
M	A	N	G	A	N	E	S	E	F	F	T	S	L
S	O	L	K	T	S	M	A	O	D	O	M	E	C
A	R	I	N	G	T	A	I	L	R	N	C	O	G
M	P	A	N	C	F	I	V	A	L	L	E	Y	N
I	C	O	A	S	P	L	A	C	R	N	A	S	I
T	L	R	V	L	O	E	D	E	S	E	R	T	D
E	I	N	A	O	H	N	R	A	T	E	C	T	D
Q	F	O	J	T	D	O	I	D	F	I	Y	P	E
U	F	I	O	S	P	T	L	C	L	R	C	I	B
A	T	S	S	P	A	S	A	S	H	O	T	I	S
R	A	O	P	R	C	D	i	S	N	N	L	B	S
T	V	R	L	L	E	N	O	N	G	N	E	E	O
Z	M	E	H	E	M	A	T	I	T	E	I	A	R
R	E	N	E	L	B	S	R	O	H	D	N	I	C

Words to Find:

SANDSTONE	QUARTZ	CROSSBEDDING	RINGTAIL	CLIFF
MANGANESE	IRON	HEMATITE.	SLOT	DOME
VALLEY	EROSION	DESERT	VARNISH	NAVAJO

Answers to the Quiz: 1. sandstone 2. Navajo Formation. 3. crossbedding. 4. desert varnish

ABOUT YOUR AUTHOR

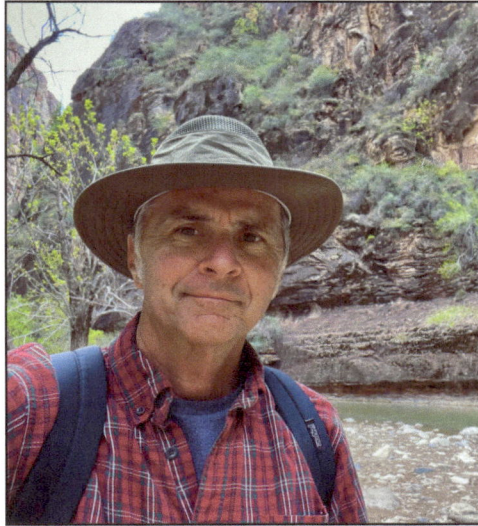

Dr. Mike Kozuch is a geologist with many years of experience studying the geology of different parts of the world. He has taught a variety of courses in geology and oceanography at the university level and is the author of several university textbooks as part of his *myScienceBlast* series. He now writes children's books on these subjects to stimulate interest in these fascinating topics. Here you see him alongside the Virgin River in Zion National Park.

www.ingramcontent.com/pod-product-compliance
Lightning Source LLC
Chambersburg PA
CBHW040812300326
41914CB00065B/1494